Rise of the Red Planet

TERRENCE PEUGH

ISBN: 1494333228
ISBN-13: 9781494333225

DEDICATION

To each and every single one of my friends,
and to Ele, my Motivation throughout the whole
writing process.

CONTENTS

ACKNOWLEDGMENTS

Mrs. Becker: My 12th grade English Teacher whom gave me this assignment in the first place

NanoWrimo: The Site that made the publishing of this book possible.

1 HUMBLE BEGINNINGS

"Set Arms! Ready! Fire!" Shouts General Wilkins as his squadron open fire on a group of every day people who had attempted to create a better life for our mining settlement and all those across the Martian surface. My name is Stephen Crow and ever since the discovery of large quantities of a new and rare mineral known as Mineral 126 or as the locals call it 'M-126' in 2097, Mining companies from all over earth have been clamoring for a piece of the action. The worth of a single pound of M-126 on the planet, is about a million credits, and with Mars projected to have over five and half trillion pounds of the mineral, there is no doubt there is money to be made from the mining of the precious mineral. After the Discovery, mining settlements sprang up all over Mars. But the problem became that the companies who were capable of retrieving the resources, had no workers willing to mine for them. With good reason too, it was an incredibly dangerous job.

With a 33% fatality rate on the job it was not the most sought after job field. The most common cause of death was being too close to the drill when it entered the ground and being swallowed by the planet. Followed by other causes such as the failure of a miner's spacesuit or poorly handled explosives. Eventually, Mining companies such as Diamond pick international, DEmix mining and Interstellar drilling all quickly realized that they needed to incentive people into taking the job. They did so by giving them and their family's safe transport to Mars, a place to live once they arrived on the planet and to top it all off an annual salary of nearly 1 million dollars. Suddenly there was a huge influx of workers clamoring to get a job at any one of these mining facility. The mining companies started to send hundreds of thousands of workers and their families to the red planet.

My father was one of the more recent waves of miners to arrive on Mars. He was one of the younger workers at the age of 38; He was clean shaved and walked with a slight limp due to an injury he suffered as a child. His brown eyes could burn a hole into your soul if you got onto his bad side. He stood at about 6' 6" in height and towered over the other workers. He lived most of his years looking towards the stars and dreaming to someday be able to escape his minimum wage job as a pencil pusher at Jennifer Mints accounting firm. So when he saw the opportunity to create a better life for his family and escape of his prison was just a signature away, he happily took it.

Initially we all hated the idea of leaving our house just so we could live in some hastily built house and fulfill my father's dream. But after some sweet talking and of course bribery we consented and the decision to move was finalized. During my father's time working under the DEmix logo five of his closest friends had direct run-ins with the company. Several were in regards to their failure of meeting their daily quota resulting in lost profit for the company and one was for being two minutes late for work. All of them resulted in the booting of them and their families off planet, and the seizing of any and all goods, including any pay they had received prior to the incident. Finally he decided enough was enough. He went to the other mining settlements to talk about breaking away from the corporations and harvesting the M-126 for themselves. The talks were successful and the separate mining settlements began to prepare for the break planet wide as to form into a single Martian Union. The preparations went on for just over a week. Then it happened.

I was outside playing in the yard with my younger brother Aaron. Aaron was an energetic boy if there ever was one. If you wound him up he could go on and on for hours. His eyes were the same as our fathers. He was not very fond of shorts and almost always wore jeans. Aaron was a boy who loved to play but not do so much of the work. His hair was of a light bronze shade.

Both of us were bouncing around the surface in the low gravity when the ship appeared. One second Aaron and I were carelessly running around doing flips, the next we were cloaked in darkness. We stopped dead in our tracks, searching around for the cause of the problem that's when we saw it. It was a monumental white ship easily the biggest and most beautiful ship I had ever seen, elegantly hovering above the spaceport. Appearing out of what seemed to be thin air. It was so big that as it hovered above the port it blocked the sun that was shining on my house and neighborhood alike. My mother, a 36 year old woman with short brown hair exited the house carrying my youngest sibling, Clementine, at the age of five to see why everything was dark. I looked to the other houses on our street and I see the expression of fear on my neighbors gathering in their own front yards. The ship slowly descended inching closer and closer to the Martian surface eventually landing on the ground with a thud. A cloud of dust echoes the thud, spreading itself across the landscape.

Rapidly approaching my neighborhood, my mother promptly grabbed my shirt, as well as Aaron's and pulled our faces into her flower covered dress. As the dust cloud began to settle back onto the ground she gradually released her grip of me. My full attention returned to the ship where I started to see men dressed in all white body suits flow out of the ship in neat orderly lines. I looked up at my mother whose face had a look of concern starting to break across it, as she watched the bulk of the men move closer and closer to where my father's Martian Union Leaders meeting was being held. She stood paralyzed watching as large clusters of men marched towards her husband. Squinting at group of soldiers she noticed that not all the soldiers were marching to the mine site, a small patch of soldiers had broken off from the pack and had begun making their way to our neighborhood. She quickly tried to hurry us all inside but I resisted, I insisted on going to check on my father. She was extremely stubborn but so was I. Eventually she folded giving me one warning. "Stephen, do not say anything, just get your father and then come directly home." Nodding I broke out into a full sprint down the dirt coated sheets of metal that were deemed to be streets supplied by the DEmix Corporation.

I arrived at the job site where the meeting was being held in record time, just under 3 minutes according to my HUD. I began weaving through a maze of barrels that my father had shown to me to be a shortcut across the job site. I began to hear loud shouting beckoning just beyond the wall of barrels that stood before me. I stood puzzling how I was supposed to get past this massive road block when I noticed a crack, not one big enough for me to squeeze through, but big enough for me to peek through to see what all of the commotion was about. I cautiously peered through the opening towards the source of the shouting. I was able to spot my father along with several other men standing beside him, and the white suit soldiers standing in a line pointing what looked like assault rifles at the group of men.

The Soldiers were very calm but extremely stern looking. Now that I was closer to them I could see the insignia on their sleeves. It was a very distinguishable image, depicting the Earth being encircled by a smaller version of the ship that had landed at the space port earlier with the words

4

Earth Defense Force arched across the bottom of the planet. I couldn't believe that they were here, the Mining companies must have reported that we were causing trouble and paid top dollar to the government to borrow the EDF to put down the growing discontent toward the corporations. Were they really that desperate? Desperate enough to call in the cavalry to fix their problem? I pushed my thoughts of why they because *why* they were here wasn't important because they *were* here and they meant business.

2 THE SITUATION ESCALATES

One of the soldiers in the front revealed a megaphone and spoke into it with a robotic coldness "Protesters, we have been ordered to put down this protest by any means necessary. Please return to work or else."

Some men began set down their plans and walk away from the table but my dad grabs one of the men frowning. "I'm sorry Cornelius, I have my family to think of and I can't afford to die today." My father frowned as he watched the men walk away letting his head hang down. But a hand slaps him on the back, it was the one of the men whom still remained; and after looking from face to face of the men who still stood beside him, each nodding their allegiance as they made eye contact with him, he smiled. He stood proudly alongside his fellow leaders looking across them as if he was speaking to each of them individually

"Do not let the military come in and uproot everything that we have worked so hard for." He called out "This is our planet! It's about time the Mining companies and their lap dogs realized it!"

The men let out a cheer beginning to make their way towards the line of EDF officers. My father made his way to the front of the protesters and began asking the soldiers why they were here.

"We have done no damage to any equipment nor have we caused any other type of mayhem, so why have you come here with guns?" My Father firmly stated. The soldiers replied with a simple cocking motion with their assault rifle and again demanded that the protesters return to work or suffer the consequences. The group of men ignored the warming and

6

continued their advance. The soldiers fearing they were losing control of the situation fired at the ground at the feet of the men attempting to scare them. This attempt backfired on the soldiers however, sending the leaders to their boiling point, faces flush with anger and rage. They slowly turned from one worker to the next nodding; silently agreeing to their next course of action. But before they could enact their plan one of the men, easily in his upper 80's, enacted a plan of his own. With cane in hand he slowly made his way to the front of the crowd. The man hobbled across the hard metal plate towards the line of soldiers who in response to the advancement stepped forward in unison and loaded their rifles aiming them directly at the old man. He paused for only a brief moment to acknowledge the action just taken by the soldiers then begins walking again. Shots ring out from a single rifle on the line, one belonging to a man no older than 19; the bullets rip into the old man's chest bringing his slow march, to an abrupt stop, weakening his knees to the point he can no longer hold his weight forcing him to place all of his weight on his can. The old man, frail and unrelenting, gripped his wound and regained his footing. He took one step forward before his can was kicked out from beneath his frail bloodied hands by the young man. The young man stood over him kicking his side so as to roll him over onto his back.

That cold hearted human lifted his foot over the wound in the older man's chest. Kicking aside the man's hands from the bullet wound; he allowed the crimson red liquid to flow freely from the wound. Deciding he hadn't proven enough of a point he stepped onto the wound, ripping cries of pain and agony from the old man's lips The young man revealed his rifle, leveling it at the old man's head; And without so much as a single glance at his target he began a relentless volley of automatic rounds into the man's skull completely desecrating his face. All while staring intently into the crowd of shocked men. My father was the first to pull himself together, he knew this was blatantly disrespectful and warranted immediate action. "Are we going to let Cletus die in vain gentlemen?!" he shouted waking them from their trance

"Hell no!" They responded back erupting forward surprising the young monster as well as the other soldiers. Several of the union leaders focused on giving Cletus some good old fashion retribution by beating the ever living crap out of his killer, while the rest began to overpower the other soldiers, bashing their heads in with wrenches or whatever blunt object happened to be the closest. They had managed to seize a small supply of weapons and ammunition, as well as a victory. But this victory was short lived. The MU leaders knew there was no going back after this point; they knew that they only had a short time to gather themselves before the Earth Defense Force sprung a counter attack on them. The Leaders stood silently looking at the mess they had just created and then to one another seeking a plan of action. Some suggested they surrender themselves to the EDF when they arrived, others suggested that they stand and fight. My father stepped forward commanding everyone's attention without a single word spoken,

"Gentlemen, what they have done today is an outright declaration of war upon us, our families, and hell even our whole planet! We must stand together now more than ever. Remember the words of Abraham Lincoln 'A house divided upon itself cannot stand.' We must not be afraid to stand together to defend what we have fought so hard to create! While I do believe that to stay and fight would be a guaranteed death certificate, I know that we must fight back. I recommend we find a base of operations, somewhere where they would never think to look, and reconvene there to formulate a strategy for removing the EDF."

The other leaders agreed. They all said that they would be in touch and began to make their ways toward the exit. But to their surprise the EDF were already set up and primed for attack. They were overwhelmed in a matter of seconds. The EDF swarmed down into the room shouting; ordering them to get on the ground. I cowered behind my wall of barrels listening to the shouting. After it had quieted down I cautiously peaked around through the crack as the men captured in the chaos were sat in a neat and orderly line.

I quickly scanned the bodies that sat across the ground. My father was not among them. *Oh thank god. Dad made it out okay* I think to myself. I had just stood up to leave when I caught sight of a large soldier entering the scene carrying a man struggling frantically to break the grip, "Dad" I gasped

"Found another one sir. This one was trying to make a break for it." Another soldier in a different outfit than the others walked over to my father's capture. The medals pinned to his shirt glistened as he walked from the warehouse. "Let him go tank this one's not going anywhere, isn't that right Cornelius?" Tank nodded and relinquished his grip on my father allowing him to fall to the ground. "Well, well look who it is." The man with the medals said, he seems far too cocky to be an ordinary soldier.

"Here I was, thinking this was just another protest that me and the boys had to put down but it's turned out to be far more than that hasn't it. Instead I have captured the leader of the Insurgency that has been a pain in my ass since I got here." I immediately realize who the man was; it's General Wilkins, head of the EDF!

"I can only say it's fitting that I am the one who puts you down. After all, this is my planet." He released his pistol from his belt and loads it.

He pointed the gun at my father; "NO" I shout from behind the barrels. But it was too late. Wilkins pulled the trigger and my father's body went limp.

3 A DARK DAY FOR THE CROWS

I had just witnessed my father be gunned down in front of me. When I had recovered from the shock tears began streaming down my face. *How could they* I thought to myself, *how could they have killed him like that?* Suddenly becoming aware of the shouting I pulled myself to my feet and peeked through the barrel wall only to meet with another eye on the other side. Both of us were startled by the presence of another person and stumbled back. The soldier began shouting and pointing towards me, taking that as my cue to leave I made a run for it. Dashing through the maze of barrels, I emerged with only the open road ahead of me plotting a clear path home. I soon came bursting through the door of my house scaring my mother half to death.

Clementine naively asks "Where's daddy?" I wanted to scream out 'They killed him! Those EDF monsters killed him in cold blood just because the rest of those hot heads were getting rowdy!'

But I didn't. Instead I lied "I couldn't find him Hun but I am sure he is alright" and picked her up giving her a hug in the process. After I put her down I walked to my mom and gave her a hug, "I need to talk to you tomorrow but right now I need some sleep." She griped me tight and kissed my cheek

"All right sweetheart, good night. I love you. " "I love you too" I echoed as I entered my room. I flopped down on my bed with the eerie creak of the metal as it bends to the force of my impact. I buried my face into the pillow and fell sleep. I was awakened by pounding on the front door and by mothers scream. The thought that the EDF had managed to follow me home to take the rest of my family from me crept into my head and I began to worry, I cautiously opened my door and made my way towards the front door. It was the white soldiers, but they were not showing any hostility. But instead offered my deceased father's mining helmet to my mom, symbolizing his untimely demise. I stood in the open and when my mom noticed me I walked towards her and embraced her. "What happened mom?" even though I already knew the answer

"Your father's dead Stephen, he fell into a hole loaded with mining explosives and was killed" she whispered. I broke down as well putting on the most convincing act I could. One of the soldiers cleared his throat grabbing our attention. He wished us the best and closed the door.

That was almost two years ago and the EDF troopers have shown no sign of leaving. In fact their Presence was being felt more and more each day. The day after the 'Accident' the EDF soldiers began to seize control of everything. Stemming from the actual mine all the way throughout our little town of 150 or so people. They claimed their reason was to better the settlement but we all knew that it was really just punishment for what had transpired with the attempted formation of the Martian Union. The bastards even had the audacity to raid our houses and steal anything that was 'potentially harmful to the occupation force'.

But in reality it was just an excuse to storm into your house, guns drawn, robbing you blind and in some cases, violating the women whose homes were being 'searched'. The longer they occupied us the more and more unrest emerged towards the EDF. Within a few days of being under EDF Control a rebellion had been formed. They eventually became known as the Martian surgies or insurgents to the EDF soldiers and 'The Sons and Daughters of Mars' to the likes of the citizens of Mining Settlement 509.

Nobody dared to reveal our support of the TSDM in the company of strangers due to the fear that the information could be relayed to the EDF who offered a small reward of immunity and food for any information regarding sympathizers or the insurgents themselves. Whenever an Insurgent group is apprehended they are paraded around the town in chains and nothing else, then promptly brought to the city square, which I might add was essentially turned into a military outpost, and publically executed. This concept was developed as a way to prevent any more of our citizens from joining the TSDM. This attempt to terrify and control the citizens was effective, and kept us all on edge throughout every aspect of our lives.

4 THE NEW GIRL

Life had been pretty tough since the EDF took over, especially at school. Pro-Earth Defense Force propaganda had been pretty much force fed to us ever since. Every day it seemed we learned of something good that they had done for the Earth on a daily basis. Like in the early 2090's when humans began to spread out to the various planets in our solar system; they managed to create a chemical mixture that could be released into the atmosphere that allowed for humans to breathe without the need for respirators.

Today's lesson was different however; it centered on the history of the moon and space exploration. I hung on every word my teacher, Mr. Garvey, was saying when there was a light knocking on the classroom door, bringing his lecturing to a complete stop.

"Well hello there, you must be... Cloee correct?" he asked as she entered the room. *Cloee,* I think, what a name for such a beautiful woman. Cloee looked to be about 5' 8", she wore a pair of jeans and a tight black t-shirt with an obscure rock band of some sort printed on it. Her hair was a shade of brown that reminded me of dry soil back on earth. Her face was completely flawless, as was the rest of her body I must say.

She responded to Mr. Garvey's question with a nod prompting him to say. "Ah, well welcome to MS 509 High School! I am your teacher from this point forward, my name is Mr. Garvey it is very nice to meet you" he said as he extended his hand to her.

Mr. Garvey is one of the few people in this hell hole I actually like. He's fairly young at the age of 27 and has plenty of enthusiasm towards what he teaches. But unlike the other teachers, he doesn't pick a side concerning the Earth Defense Force and The Sons & Daughters of Mars fighting. With his jet black hair and plaid button up shirt, kakis and a varying crazy tie each day making him a very hard man to miss. Cloee extended her arm to meet his and shook. Mr. Garvey pointed her towards the open desk diagonal from mine. She quietly made her way to the seat and sank into the chair. Mr. Garvey then began his lecture from where he had left off at before Cloee had arrived.

"So as we all know, the first Human Settlement outside of the earth's atmosphere was the moon. In what year was it colonized? Cloee do you know?"

She immediately snapped to attention at the mention of her name, "Wait, what about years and the moon?"

There were chuckles throughout the room myself included; Mr. Garvey repeated the question and Cloee responded "Um wasn't the moons first colony established in 2047 with the U.S.S. Maverick?"

"Very good! And do you know the man who is credited as the man who led the colony for almost 34 years until he was overthrown by Revolutionaries?"

"Well actually it wasn't a man who led the colony, it was a woman named Susanne Rider. She created the first lunar gravity omission device to make it just like living on earth…at least gravity wise; then she was able to figure out the best way to farm crops in the lunar soil." The whole class fell silent and stared at the new girl in disbelief. Mr. Garvey just smirked and said "Nicely done miss…."

"Steele" she replied "Cloee Steele"

5 THE MOON FALLS

I think I am in love. I know it may sound corny to believe in love at first sight but with this was for real, she was an absolute knock out. She had jet black hair and gorgeous brown eyes that coincided perfectly with her hair. She seemed a little shy but insanely brilliant on the subject history Her lips curled ever so slightly when ever she smiled, giving a slight glimpse at her perfectly white teeth. She was wearing an old Mineshaft band T-shirt with a pair of wicks jeans on, covered on the bottom in the crimson martian soil. She had a modest bust for her size and was stunning from head to toe. I couldn't even take my eyes off of her as she stood sheepishly in the door.

Mr. Garvey is the one who rips me from my thoughts back to class by calling on me. "STEPHEN! THINK FAST!" He says whipping an Anti-gravity disc at me. I couldn't react fast enough and got straight up hit in the head with the Disc. I fell out of my seat to an uproar of laughter from my classmates, and I must admit I was laughing as well. After I picked myself up off the ground I looked to Mr. Garvey who was beside himself with laughter. He looked as if he was about to keel over from laughing so hard.

"Maybe that will teach you to not pay attention in my class Mr. Crow!" he says hardly able to speak from the lack of breath.

"It sure has, but it more importantly taught me to bring a helmet with me to school!" I respond rubbing my head, which garnered even more laughter from my peers including a snicker from Cloee that I caught out of the corner of my eye. Once he had composed himself Mr. Garvey asked "Now that you are back in reality would you care to answer the questions I asked you?"

"And what questions would that be?" I respond

"What year did the EDF successfully defeat the lunar revolutionaries?"

"The lunar revolution began in the year 2081 with the overthrow of Susanne Rider"

"Ah congratulations you can add correctly Stephen. But do you know what began the revolutions?" Mr. Garvey inquires.

I take my seat and reply "Ya know I think I'll let you take this one Mr. G because I have no clue."

"Very well Stephen, but instead of having me strain my *enormous* brain why not call on one of your peers." He says with a small smirk on his face and a wink. Taking his not so subtle hint I say "Very well sir, hmmm Cloee do you know what the answer is?"

"Huh?" she replies snapping back into reality "What did you say? Sorry I missed the questions…."

"Miss Steele I suggest that you pay attention in class otherwise unless you want to follow in the footsteps of your classmate, Stephen and take a disc to the side of your head." Mr. Garvey replies from behind his desk grabbing the disc that hit me earlier

"Which I mean if that's your wish I will happily oblige you, I need all the practice I can get for my anti-grav disc golf tournament coming up"

He says standing up and flexing his muscles and throwing arm which commands a laugh from the class. Starting to glow red Cloee composes herself and replies "No thank you, I am more than content with not getting hit in the head with a Frisbee, so what was the question I was asked?" Leaning forward towards her I whisper it to her

"What began the lunar revolutions in 2081." Immediately following the completion of my sentence I receive another thwack, this time on the forehead, by the same disc as earlier "OW! What was that for?" I ask while I rub the front of my head

"No giving hints Mr. Crow, you know that!" Mr. Garvey responds "Plus I just wanted to hit you with the disc again."

"Well that's fair I suppose. But it still hurt." I say as I throw the disc back to him.

Catching the disc he replied "And it was also funny so it doesn't matter, anyway Cloee what is the answer to the question."

She appeared to be in a state of deep thought when she was asked. After a minute had passed she looked to Mr. Garvey, who was patiently waiting for her to answer, and said

"The lunar revolution began in 2081 because of the lack of representation the colonists were getting in the Worlds Congress and the oppressive treatment of the planarians by the EDF. They were led by Constantine Philips and carried out raids on several supply depots across the lunar surface."

"Very nice! Good job Cloee, Stephen maybe if you paid half as much attention as her you could avoid getting hit in the head by a disc." Mr. Garvey says with a little wink at the end to confirm to me that he is just giving me a hard time

"But allow me to elaborate on this subject just a little bit more. The uprisings were put down in less than a year by EDF troopers. The last effort made by the revolutionaries to break free of their chains started with a massive underground meeting. It was said that over half of the population of the moon showed up to watch Constantine speak. He presented his dream for a unified moon at the iconic location of the first moon landing from the 1960s. His dream was to create a separate entity from earth which could easily become a self - sustaining nation free of any oppressive government. His speech garnered unimaginable applause and praise.

Following his rally he positioned all of the rebels around the EDF main encampment and attacked. The EDF were caught completely off guard by this attack and were all but annihilated. For a brief moment it appeared as though they had won their independence. Then over the horizon appeared a Gigantoromundus-" The word alone would have gotten a few laughs out of the entranced class but his arm motions were what did it for me, after the laughter quieted down he continued "-white-ship that had enough fire-power to destroy an entire planet. Its appearance alone made the rebels freeze in their tracks, killing all of the celebration that had begun around their victory. The ship, The EDF extremity, didn't land but instead hovered about the scene of carnage and destruction. Soldiers began to drop out of the Extremity at an alarming rate.

Seeing his men being killed in front of him must have been too much for Constantine he soon called for an end to the fighting. The leader of the ground EDF troopers, Corporal Wilkinson, agreed and signed the Wilkinson-Phillips treaty which stated that Constantine would enter EDF

Custody if and only if all of his rebels were to be acquitted for their crimes and allowed to return to their families. Wilkins agreed and the treaty was signed!"

The bell sounds signaling the end of class. Hands begin to fly as my classmates begin to pack up their bags but mine are slow to pack as I carefully watch Cloee as she begins to pack up her things. Sensing that someone was watching her she whips around and I have to very abruptly look to my bag as I begin to literally shovel my work into my book bag. She very lightly giggles and turns back around to finish packing her things. She begins to stand to leave, and not wanting to lose my opportunity to talk to her I swiftly sweep the rest of my papers into my backpack and stand up next to her. The two of us walked out of class into the hall wishing Mr. Garvey a good rest of the day. We walked down the halls in complete silence for a time but I finally finally broke the silence between the two of us. "I don't think we have been formally introduced, the name formally designated to me via my parental units has been set forth on the day of my birth as being Stephen Crow."

She giggled at my over dramatization of my name and sheepishly smiles, she carefully slid her sun touched brown hair out of her face that was even more breathtaking up close than it was from far away. Our eyes met

"Well hello there Stephen, the name formally designated to me by my parents on the date of my birth is Cloee Steele." Extending her arm to me "It's nice to meet you."

"It's nice to meet you as well" I say meeting her hand, gripping it and granting it a firm shake. Our eyes meet again but this time we each become transfixed on one another. We release each other's hands after realizing that we each had been holding hands longer than either of us had intended. We both stood staring at one another in the hall until Aaron comes bursting through the doors frantically running towards us. He was completely out of breath by the time he reached us, he could hardly be considered audible "Just calm down there junior. What's wrong?" I genuinely believed he was about to keel over from hyperventilating so much. He eventually stood up straight trying to catch his breath, and lets out "Its Mom."

6 A CLOSE CALL

My heart sinks as those words collided against my eardrums.

"Aaron, what is wrong with mom?"

"I'll tell you in the ATMV."

"The ATMV?" The three of us- Cloee, Aaron and myself- rush out the doors and there sitting just in front of the school is my father's old ATMV

"Aaron how the heck did yo- never mind, I don't want to know. Just get in, we don't have much time." We all pile into the ATMV; once everyone was strapped in I floored it, inevitably peeling out and sending pebbles flying into the air behind us.

"Alright Aaron spill it, what happened to mom."

"Well it all started when three EDF troopers stormed the house, they never once told us what it was they were looking for. They just kicked in the door and started shouting. Their guns were drawn as if our family posed some kind of a threat to them, then they noticed the picture of mom and dad on their honeymoon on the table. One of them grabbed hold of mom, showed her the picture and questioned her if she knew the man in the picture. Of course she nodded saying that the man in the picture was her husband. They then accused her of being one of the masterminds other than dad for the rebellions that were starting to take place all over the planet. Then they just took her, I think they were taking her to the line."

"THE LINE?!" I slam on the gas pedal watching as the RPM climbs rapidly then drops as the gear changes. Cloee can hardly believe what

Aaron has said either,

"Why would the EDF take your mom? Aren't they supposed to be keeping the peace anyway? Why even bother storming a civilian house when absolutely no probable cause is present? It just doesn't make sense to me."

"Because these men are not the hero's they may be portrayed as on Earth, if anything the rebels should be the heroes and the EDF should be the terrorists." I answer, never removing my eyes from the road. I peer into the rear view mirror to see if there is anybody following us. As I check the mirror I catch sight of Aaron looking at me mischievously, hinting that he knows that I have a crush on this girl.

"So Stephen, who is your lady friend?" He says with a grin across his face

"Hi, I'm Cloee. I just moved here because of my father's work. He just got hired as an EDF soldier and as a first assignment he was sent here to mars." I take my eyes of the road for a brief second to take a look at her and think *Of course. The one time I find someone who might be interested in me; she ends up having a father who is a part of the very organization that is threatening to take another member of my family from me.* Aaron and I lock eyes in the rear view for a split second, but it was plenty long for both of us to sense the doubt growing from her statement and whether or not she can be trusted, but not wanting to make it any more obvious that neither of us were comfortable by this new information he extends his hand.

"Well it's nice to meet you Cloee."

"It's nice to meet you as well" The two shake hands and return to facing forward. I start to slow down as I enter my neighborhood and pull into my driveway. "Out." Aaron is bewildered by this and responds "Why would you kick Cloee out? This isn't even her house it's ours."

"I'm not kicking her out, I am kicking you out. Now go get Clementine and go over to one of our neighbor's houses. Tell them I sent you and explain the situation alright?"

He began to open his mouth but I shot him a look that said *if you so much as say a word I will forcibly throw you out of the car*. He gets the message loud and Clear

"Just be careful Stephen we can't afford to lose you too." I nod. He shuts the door of the ATMV and runs inside.

I slide the ATMV into reverse and whip out of our driveway, and speed off towards the 'Earth Defense Force Forward Operating Base' or EDF FOB for short, where 'The Line' is located. It is located on the left side of what use to be our town square prior to the invasion. When they took over the square they surrounded it with wire fencing and set up guard towers in an effort to make sure we weren't communicating amongst one another. They kept a tight grip on the square only allowing the use of it from 12:00 to 3:00 for shopping. This base was not impenetrable during the other 21 hours of the day however; there was a single dead spot where no one but the person on the other side of the fence could see you. There you could climb the fence, gaining access to the rest of the base.

As for 'The Line' it was a trench that had been dug out so as to catch the dead as they fall forward, the trench itself has an incinerator of sorts attached to it. This was done so that no one had to handle the dead bodies, a task that had proved too demoralizing for the vast majority of the soldiers. The final touch that was added so as not to accidently injury an innocent civilian or soldier, if such a thing even existed here on mars, was a large concrete slab that's intended goal was to stop the bullets from the EDF rifles from hurting anyone who may be stupid or unfortunate enough to enter area behind the firing zone. The square is about 10 minutes from my house so we arrived in no time. Just before getting too close to the town I pulled the ATMV to the side of the road and backed into a space between two rocks so as to hide the vehicle from the EDF or anybody else for that matter.

Cloee initially questioned my rational at doing this but figured it was a better question for a different time she followed close behind me as I scaled the fence and crept around the building to a small spot where two wooden crates were stacked atop each other. Peering around the crates I spot several of my neighbors being filed out of an EDF transport van that had clearly been on the planet for a while. The once bright Silver rims as well as the bottom edge of the van were now stained red from the planet's soil. In hindsight perhaps bringing an all-white van to a place that has the nickname 'The Red Planet' wasn't really the best idea. Cloee nudged me anxiously "Hey what's going on over there?"

"They are unloading the rebels from the van and starting to line them up on the ground." The last of victims exits the van and I quickly survey the line looking for my mother but she is nowhere to be found. Thoroughly confused I lean back against the wall and slowly slide down it. Plumping on the ground staring down towards my shoes stained red by the ground. *Where is she! This isn't the sort of thing Aaron would joke about. He knows how I feel about this kind of thing. So there is no way he would lie to me about this, especially coming into the school as winded as he was.*

"What's wrong Stephen? Did you see your mom?"

"No I didn't. That's the problem…" Then I had a stroke of genius. "Hey, has your father ever talked about other places they would take the prisoners of war? Because up until now it seemed like they were only taking them here to be killed."

She stood just above me scratching her head trying to remember.

When it came to her, her eyes widened her hand shot up with her index finger pointed towards the sky "I've got it!" she exclaimed, immediately clasping her hands over her mouth aware of what she has just done. I quickly stand up and peek around the corner to make sure we weren't heard. As I had feared the soldiers were looking in our direction, *Yep. We're boned* I thought. The two soldiers started to cautiously walk over to our position guns drawn and the safety off. I shot Cloee a quick look; she mouths the words 'I'm sorry'. I mouth back 'its ok.' I hear their boots getting closer and closer with every step. I squeezed my eyes shut and waited for the end. I felt my heart thud against my ribs as I sat in fear what they would do if they found us. Cloee slid her hand into mine and refused to let go, in fact she began to squeeze the life out it, out of fear no doubt, so I squeeze back. I heard the steps come just before the crates Cloee and I were cowering behind then suddenly; they stop, and begin to quickly move away from us. "Oh sh-…." Was all I was able to hear before The EDF van whipped around the corner and smashed through the front gate of the compound, setting off the bases vast array of alarm systems.

All of the sirens began mixing together into one ridiculously irritating sound. Cloee and I cover our ears trying to protect them from the mosh pit of sounds echoing across the base. I signal to Cloee that this was our opportunity to leave, she nods and made a run for the fence. I cup my hands to help her up and over the fence. As soon as she was over I began the climb as well, at the top of the fence I glanced back towards the base to see what was happening. I could see that the few troops that were still on the base were organizing themselves to chase after the escaped rebels. They piled into their specialized ATMVs, and started to give chase. Surveying the parts of the base away from the barracks and Garage I noticed what could have only been the two soldiers that were encroaching on Cloee and I's hiding place, thus inadvertently allowing the rebels to make off with EDF equipment as well as their own lives. The dynamic duo was being yelled at by General Wilkins for the actions (Or lack of actions I suppose) earlier. He was screaming so loud it was clearly audible from where I was on the fence, clear across the base.   "How the Hell did you two dog gone ninnies let them get away! Your task was not a difficult one. Point the rifle at the insurgents, and pull the trigger. THAT SIMPLE! Now they have an EDF Van, which no doubt has weapons inside of it, so you two just gave those surgies the means with which to fight back. This kind of thing is considered an act of treason you are aware correct?"

The two soldiers respond but they were not being anywhere near as loud as the General so I couldn't make out exactly what they said. "Good! Now go and find those rebels, otherwise you'll be the one on that firing line!" He started to walk away but after about thirty or so steps he turned around.

He pulled a beam pistol from his belt and shot the man on the left point blank removing his strength and causing the man to fall to the ground in a heap. The remaining soldier was shocked and immediately sent scrambling to get into an EDF assault ATMV. I felt no remorse for that man.

7 REUNION

Cloee and I hurried along the fence to where I had hidden the ATMV when I heard what sounded like an engine making its way towards us. We were only a few yards from my ATMV. I knew that if we were discovered then we were sure be the next victims of the line.

"Come on, we have to make a run for it. They are coming." I shout to Cloee, who acknowledged me by making a break for it, with myself not far behind her. Quickly glancing down the road as we cross, I see a cloud of dust inching closer and closer to us I picked up the pace keeping just behind Cloee. She was the first of the two of us to make it to the ATMV with myself not far behind her. I threw open the door of the ATMV and slammed the keys into the ignition while she made her way into my passenger seat. With both doors shut I started the ATMV, shifted in reverse and pulled back into my little crevasse because I knew that I had no chance in out running an EDF assault ATMV. As I pull back into the crevasse further and further the more I realize that it was a normal one. The roof of it began to close above me blocking the sun from entering effectively becoming a cave. After I realized that I was far enough to be completely hidden from the road I turn off the ATMV and we sit in silence holding our breath. The set of head lights I saw on the road turns into the cave the driving at an alarming fast rate. The lights sweep across the cave and my heart starts beating rapidly. Questions begin to fly throughout my head. *Did the EDF see us run into the cave? What are they going to do with us if we get caught?* Luckily my questions need not be answered because the lights come to a stop then disappear into the darkness. Staring at the last place I saw the head lights pondering where they lights had gone? "I'll be right back; I need to go check something out real quick." I whisper to Cloee who still has yet to take a breath. Her forehead wrinkles with worry "What in god's name do you need to check? Those headlights could belong to the EDF soldiers that are hunting for the rebels at this very moment! Just let them b-" She is silenced by the sounds of sirens echoing through the mouth of the cave as a herd of EDF ATMV's sped past the opening of the cave.

I look at Cloee with smirk that signifies my victory in the matter "See those were the EDF, whatever or whoever just entered the cave isn't EDF. If it's anybody I am willing to bet it was the rebels entering the cave. So I am going to go over there and make sure that those are rebels, because if anyone knows where my mom is, it will be them." "OH, speaking of your mom I think I know where she might be!" That's right the whole reason we were in this situation right now was because of Cloee's outburst at the base. "Hold that thought! Look!" I say motioning towards the lights that have reappeared ahead of us in the cave. I can tell from the reflection of the lights off the walls of the cave I was correct in assuming that the lights belonged to the rebels. They shone onto the wall and poured across the front of the van, revealing the EDF insignia on the passenger side door. Men and women begin to climb out of the rear of the van and start to congregate by the passenger door. A solemn man hangs out of the passenger door standing on the edge of the door; He commands the attention of the others with no words at all. His sheer presence is enough to silence the crowd and garner their undivided attention. The man was no older than 43 from the looks of his body, but his hair and face were aged terribly from stress. His hair predominately retained a milky white appearance to it; the hair itself wasn't very long perhaps only down to the middle of his forehead with no effort made to style it. His face showed the scars of battle with a large scar making its way across his face and through his left eye which bears an eye patch. A beard was beginning to take shape on the lower half of his face. There was no doubt in my mind why this man was the leader of the rebels. He looked almost to have an aura of control and premise. I climbed from our ATMV and make my approach towards the rebels. One of the rebels spots us as we approach and quietly leans over to tell the head rebel, the man comes to a stop and looks to Cloee and I after hearing this information. He glances up over the small group, without saying a word he steps down from the van and makes his way through the crowd. The group slowly turns to the rear to see what could have caused their leader to stop his speech. When they make eye contact with us they all take a collective gasp, separating to allow the man to pass through. He reaches me and stares into my eyes with his dark brown eyes. They seemed familiar to me; the man began to cry softly. "Son." Embracing me and holding tight.

8 THE SONS AND DAUGHTERS LIVE ON.

I couldn't believe it. My father, whom I have spent the past couple of years grieving over, is here, standing before me hugging me so tightly I can hardly breathe. We stand like this for only a few minutes but it felt like an eternity to me. Having somebody return from the dead can do that, especially if it was someone who you watched, or at least thought you had, get killed. When the embrace was over my father and I look to each other beaming to see each other. He motions for Cloee to join us and she begins to walk over to us. My father extends his arm to her as soon as she arrives "Hi I'm Stephens's father, Cornelius Crow. But you can call me Cornelius. Everyone does around here." He says widening his arms to encompass the entire cave. "It's nice to meet you Mr. Crow, I mean Cornelius. My name is Cloee Steele." She grabs his hand and shakes it. My father then turns his attention to me

"So how did you come about being in this cave?"

"Well actually dad, we are the reason you folks are all still alive and not a pile of ashes in the incinerator pit."

"What do you mean" he inquires

Well Cloee and I were at the base when you were being lined up. Cloee shouted on accident and the soldiers put in charge of killing you came over to investigate the noise. Thus buying you time to not only escape but to steal an EDF Van, destroy the main gate to the FOB and send General Wilkins into a fit of rage on the soldiers that were supposed to kill you. The monster even killed one of them because of it, shot him point blank through his chest."

He stares into my eyes and motions for the two of us to follow him. We walked to the group of rebels that were still standing waiting for my father to return. Cloee and I stop in the crowd but my father continues, he climbs the platform he had been on before we had arrived. He stared into the crowd moving his eyes from face to face of the rebels. Then restarted his speech from the point he left off at when Cloee and I had shown up.

"Now where was I…Ah yes I remember. We are only a few big moves before these bastards are out of our hair for good. Being captured and having our encampment destroyed dealt a serious blow for the resistance. But I have not given up hope! Because as long as I live and breathe, I will make certain that they don't have a single easy day. That they will pay for occupying our planet! They will pay for killing our neighbors. They will pay for killing our family! Let us rise above the EDF as one and free our planet!"

The group erupts into applause and cheers as he stands over them all smiling. He throws his fist into the air. The group follows suit and throw their fists into the air as well. I look to my father who is grinning at his work; even Cloee had become engrossed in the situation throwing her fist into the air and cheering for my father. I joined in the celebration just as it was at its peak; my father placed a single finger over his mouth signaling for the crowd to quiet down. They obeyed and quieted down to a dull hum.

"Tank, I need you to train the troops and get them ready to fight because they look like they could use a little bit of help getting them to fighting shape."

"Copy that." Boomed Tank, I understand exactly why this man has the nickname 'Tank' I mean he is incredibly ripped and seems like he is the kind of person who could break your back with one hand!

"Lopez, I am going to need you to work on this EDF transport van making sure they don't have a tracker on it somewhere as well as repurposing it for our uses."

"You got it Cornelius." Lopez replied giving him a thumbs up sign, this man was the complete opposite of tank. He actually looked fairly average. With his shaved black hair, and oil covered fingers I could tell he was more into cars then he was into lifting weights in his free time.

"Lilly, I want you to tend to any of the injuries that any of us have suffered."

"Yep no prob." She says with a smile. Lilly looked like the youngest out of the rebels. I'd guesstimate that her age was around 24. Her jet black hair was cut short just above her eyebrows; it was carefully split into two curves over her forehead. She seemed like she could handle a fight should one present itself.

"Last but not least, Stephen I want you to return home to take care of your mother, brother and sister." This I was not expecting, I would have thought that my father would want me to follow in his footsteps and help him fight the EDF I certainly had a cause to do it. So I Protested

"I can help you though!"

"No Stephen. This isn't your fight, you are still too young. You have your whole life ahead of you. My goal for you, and everyone else on this planet for that matter, is to have a life free of outside intervention."

"But…."

"No buts. You are not to participate in anything we do. Understood?"

"Yes sir."

"Now as for the rest of you, I want you to get this place ship shape! Because I think this cave is as good a place as any to set up shop in."

"Good. Now…. you all have your assignments we have a lot of work to get done. But with any luck everything will go according to plan and this planet will be free by next month!" The Crowd once more erupts into applause, throwing their fists into the air. My father replies in kind by putting his fist into the air. After the crowd had dispersed to complete their various tasks he walked up to me with a purpose. "I'm sorry for calling you out like that. But I just don't want to lose you okay. This isn't and never will be your fight okay?" "This most certainly is my fight." I respond "If you had your father whom you thought was dead return from the grave then proceed to tell you to essentially get lost you be upset too."

A look of guilt spreads across his face as he recalls the day he 'died' leaving my mother alone to raise three children and mentally scaring me with the image of him getting shot. "Stephen…I did what I did for reasons." "Like what dad?" I am practically screaming at him. "Did you want to leave mom? Or perhaps you did it to abandon all of your kids? Hell maybe you wanted to hurt everyone who has ever know or cared for you!" I grab Cloee who is still standing beside me and lead her back to the ATMV.

"Maybe I knew that if the EDF would send in this big of a force to quiet our attempt to overthrow them, they wouldn't hesitate to attack our families, and if they did maybe I didn't want my family getting severely hurt as a result of my actions towards the EDF."

"Oh by the way, Mom was arrested by the EDF strictly because she was your wife and is being held who knows where; so good job protecting the family dad."

"What?" he says in a rough voice

I lead Cloee to her side of the ATMV and open the door for her before I continue. "Yeah she was taken today that's why we were at the base today, we were looking for mom. But instead of saving her we accidently saved all of you, but we didn't see her in your group so there must be some other place that they keep the prisoners. That's why Cloee screamed because she thought of the place and got the soldiers attention."

My father's forehead creases as he processes the information I have just given to him. He motions to Cloee who is quietly sitting in the ATMV waiting for me to join her. She nods, exits the vehicle, and joining me at my side. "Alright Miss Cloee, my wife has been taken by those EDF scumbags. Now if you would be so kind as to tell me where they might be keeping her I would be very appreciative." Cloee acts as though she were a deer in the headlights of an oncoming car when she is in my father's presence, she only acts like this for a split second before she snaps back into reality.

"If they aren't killing them all at the line then that would mean they have an encampment somewhere no one would ever be able to see. I had heard my father talking about it to a couple of friends about the EDF beefing up security at an internment camp somewhere northwest of the Forward Operations Base. So with any luck, that is where they are keeping your wife." My father looks thoughtfully into Cloee's eyes scanning her facial expression for any more information she may be able to offer but she remains silent. He grabs her shoulder and says in a sincere voice "Thank you for your help." Then releases his grip on her and turns to face the group. He begins to speak to the crowd but realizes to get their attention he needs to be on top of something. So he quickly climbs onto the hood of his old ATMV and begins his speech. "Alright Folks there has been a change of plans. The first thing we are going to do is break into an EDF internment camp and rescues our Family and friends. I need two of you to go and locate the camp itself, that way we can understand what we are going to be dealing with when we storm the place. So do we have any volunteers?" He slowly moves his head from right to left scanning the crowd for hands but no one is raising theirs. "I Volunteer." I shout and hold up my hand "I do too." Cloee says as her hand shoots up to join mine no longer making it the only one that was up.

My father's forehead creases again, but due to the lack of volunteers he is forced to accept it. "Alright so it looks like Stephen and Cloee will be the ones scouting out the facility whilst we prepare for the attack." He climbs down from the ATMV and rebels scatter all across the cave returning to their initial assignments. My father pulls us aside and tells us "This is going to be extremely dangerous mission; if either of you two get caught they will probably throw you into the very camp you are investigating, or the more likely of the two options, you two are put on the line and killed." His words hit me like a brick but I know the reality of my actions and what they may cause, everyone in this cave knows exactly what the EDF are capable of doing to them and Cloee and I are no different. We know exactly what is at stake.

After my Father had given us instructions in regards to the mission we had accepted. We had given him a quick nod of the head and climbed into the ATMV. We drove slowly through the cave which had become busy with activity in preparations for Operation: 'Martian Freedom' as my father had dubbed it. All it entailed was leading a small group of rebels disguised as EDF Soldiers onto an EDF attack ship and take it over. From there the plan was to then use said aircraft to stage a raid on the EDF internment camp where my mother and countless others no doubt, are being held. Then from there he plans to arm the rebels he has acquired and lead an assault on the EDF Mother ship. Now I never said my father ambitious but this is a little too circumstantial for my liking. I mean we literally need everything to go perfectly for us to even remotely dream of taking over that mother ship! This all just seems like one big old crap shoot. But it's the best plan we've got so we might as well stick to it, right?

Cloee was right about the Internment camp only it wasn't really all that far from the FOB, only about 20 kilometers away at most. When Cloee and I arrived just under half a Kilometer from the camp I pulled the ATMV onto the side of the road and do my best to try to hide it from sight. Once I am satisfied I signal to Cloee that it's time to go scope out the camp. We climb to the top of a hill overlooking the camp and lie down on our stomachs. I pull out a pair of binoculars and start to survey the area grunting at what I see.

"What do you see Stephen?"

"A whole lot of problems Cloee, a whole lot of problems." I respond pulling my eyes away from the binoculars and turning to her.

"What do you mean 'a whole lot of problems'?"

"Take a look for yourself." She takes the binoculars and starts look around the area witnessing what I have already seen. Hundreds of soldiers armed to the teeth with technology that far over shadows our meager supply of weapons. Not to mention the Guard Towers at each corner of the camp with fully automatic rifles by the look of it, the 5 meter tall fence that I can only assume was electrified and only one way in or out. Cloee sighs as she takes her eyes away from the binoculars "Your right there quite a few things to overcome with this place."

"Yeah, I think the hardest part of this whole plan besides getting in there in the first place, is going to be getting out alive I mean the only way you get in and out of there is if you're an EDF personnel!"

I lay flustered about the execution of this jailbreak then I realized we had all the resources to pull it off now. "I've got it! We need to go back to the cave now. I know exactly how we are going to break into that camp."

"How"

"Two words: Trojan horse."

"How the hell is that relevant to our situation right now Stephen?"

"Come on I'll tell you on the way." We both crawled backwards a ways, got to our feet and made our way back the ATMV. Once we were both in I pulled out of the hiding spot we had chosen and pulled onto the road towards the cave. "Alright Stephen spill it. What's this fantastic plan of yours eh?"

"Ok so you know how my dad and the others managed to make off with that EDF Transport ATMV?"

"Yeah but what does that have to do with breaking into an internment camp?"

"Ok try to follow me on this. Ok so like you said earlier the only way in and out of that camp is if you are an EDF trooper, or what they perceive EDF troopers. So what I propose is that we pretend to be preforming a Prisoner transfer to FOB for execution, but instead we just take them to the cave where it's safe."

"Hmm not a bad plan at all Crow, but how do you plan to get the Uniforms?"

"I am assuming that the EDF Transport ATMV probably has uniforms of some sort stashed inside of it somewhere. We just need to get there before they completely dismantle the thing." Cloee nods in agreement "So if I were you I would step on it before it's too late." I smile, pressing my foot against the metal pedal pressing Cloee and I backwards into our seats as the speed increases. It only takes us half an hour to get back to the base. We had gotten there just in time because they were about to paint over the EDF insignia when we skidded into the cave garnering everyone's attention and almost hitting a few in the process. I leapt from ATMV after it was parked and informed my father of the information we had gained from our mission. A look of frustration crossed his face "Damn, so it looks like the EDF has gotten a scared of us. But how are we supposed to get into the base if it is that well defended?" I explain to him about the plan I had concocted. Operation: Martian Horse as I had dubbed it in my head. He listens intently throughout my entire explanation of my scheme. Nodding occasionally to let me know he understands everything I am saying. When I finish he pauses for a moment to think over everything I have just told him. I stand back looking at his un-breaking face waiting for his response. "That has to be one of the craziest and circumstantial plans I have ever heard...... Let's do it." He says smiling and patting me on the back. Looking away from me he calls for everyone to gather in the center of the cave. Within moments bodies are huddled around my father. "Well everyone, we have a plan of attack for the camp where our neighbors are being held. This plan was devised thanks to the intelligence gathered by Stephen and Cloee." He motions his arm towards the two of us and a quiet round of applause starts up briefly but dies out quickly. "And what this plan entails is...you know what I think I'll let the master mind enlighten you all, Stephen will you please step forward and explain your plan?" Crowd parted so as to allow me passage to the center of the circle, when I arrive at the center I take a quick spin and take in the amount of people staring at me. Hanging on my every word even before I speak, Taking a deep breath I begin to detail to everyone my plan. Noting at the beginning of the presentation the name of my plan which does manage to garner a few laughs but those are short lived as others shush them. When I complete the explanation my plan people are looking back and forth to their fellow rebels exchanging nervous murmurs. Finally Tank steps forward "I'm in little crow, I for one am getting a little tired of running from these EDF clowns. Let's get in there and show'em what we are capable of doing whose with me?" "I'm wit you Tank" Lopez says as he exits the massive crowd. Tank extends his fist to Lopez who taps the fist in return with his own clenched fist. "Oh what the hell I'm in too." Lilly beckons as she exits the crowd "Let's get these EDF fuck wads." Soon everyone is starting to get behind the plan and soon I have the overwhelming support of all. My father grabs my shoulder nods approvingly and leans down to my ear and says so that I can hear him "Looks like your plan is a go boy-oh."

11 THE PIECES FALL INTO PLACE

The days seemed to fly by after that. With everyone firing on all cylinders trying to get things ready to go Cloee and I had plenty to do. Between spying on the camp to try to work out the schedule of operations, to helping out Lopez with setting up the transport van they 'borrowed' from the EDF making it more fit for our needs being careful to preserve the outward appearance. Cloee was able to score two blank EDF IDs from her dad that she claimed were for a school project. What took a week to accomplish seemed like it was completed in only a few days.

When all of the prep work was done my father called a meeting to go over the finalized mission plans. "Alright everybody, tomorrow's the big day. And allow me to be the first to say how proud I am of everybody for the amount of work they have put into this mission. Now I am not gonna sugar coat this. Once we get in there shit is going to get real. Some of you may not even make it out alive. But if we play this smart then we won't need to worry about that. Now allow me to have my son Stephen remind everyone of 'Operation: Martian horse' as he dubbed it."

He steps back nodding to me that it is my turn to speak I nod back and step forward, attracting the attention away from my father. I take a deep breath "Alright folks, here's the plan in case you have forgotten. Phase one is to first and foremost get into the camp, security has remained the same as far as Cloee and I can tell at the base so most if not all the intel we have gathered thus far will be valid. Phase one will require the Two EDF Suits we have as well as the Fake Identification cards, I think it would be best for Lopez and my father to drive the transport van. Do either of you object to that?" They both look to each other and then turn back to me shaking their heads. "Okay good, you two will be informing the gate keeper that we are preforming the routine prisoner transport to the prison mine a few miles out. As for the verification codes that will authenticate our claims Lilly has obtained several from the several drunken EDF soldiers while they were at the bar. Phase two is locating the prisoners and will involve the same two people as phase one. After we get into the base we should be led by several troopers to the holding cells. As soon as you two have entered the area where the prisoners are being held; you need to knock out the two escorts discreetly, preferably hiding the bodies in the holding cell. If there should happen to be more than a couple of guards, be careful about how you deal with them the last thing we need is that entire base knowing we are there."

"What do you mean 'we'? How many people are going to be in that ATMV?" Tank asks as he steps towards the center of the circle

"I mean our little group: You, Me, Lopez, Dad, Lilly, and Cloee. We can't afford to risk just two of us going in there and getting killed. We are going to need the option of back-up in case things go south."

"But how are we going to get all of us onto the base without being found?" Lilly points out. "Well that's simple Lilly, we don't let them see us" the voice of Lopez responds lacking a body to claim ownership of it. Everyone looks around trying to find him to no avail when he reveals himself at the center of the circle holding a blanket. "Allow me to introduce to all of you the latest innovation in Insurgency tech; the Interior Cloaking Device or ICD for short" I say excitedly

43

"What it does is it takes the light beams that reflect off of objects and bends them around the blanket making beneath it essentially invisible. Lopez himself came up with the idea for it, I just helped make it a reality." His face lit up like beacon beaming with pride over his invention that had captured everyone's attention.

"So we will be using the ICD's as a way to sneak the rest of you onto the base in the back of the Transport ATMV." I continue

"As for phase three, this one will be the toughest part of my plan, Because from what Cloee and I have gathered one transport is not going to be anywhere near enough to get all of the prisoners out of the camp. So what I propose because of the loading procedure the base has with backing their transports up almost directly to the cells where the prisoners are held, is that once the truck is in position two of us will break off from the group 'borrowing' the uniforms from the soldiers taken out by my father and Lopez. Then from there the two who took the uniforms will need to get their hands on a second transport and load it with any remaining prisoners that would fit in the first. Once we have everyone loaded up and ready to go we'll need to get out of the base obviously. With two ATMV's instead of the one we entered with, questions are bound to be raised as to our intensions so we will need a distraction, and tank that's where you come in." "What do you need me to do little crow?"

"You've worked with explosives before right?" a devious grin spreads across tanks face answering my question before he even speaks "Once or twice, yes."

"Good. You are going to need to set up explosives by the watch tower, vehicle bay and in the holding cells. Will you be able to do that?"

"Not a problem. I'll blow the whole place sky high."

"I like the enthusiasm but we don't want to die so you are going to have to connect them to a detonator; So as to prevent them from blowing prematurely."

"Can do, am I doing this solo or am I going to have a little help to watch my back?"

"I don't see why not." Tank nods his approval, signaling me to continue with the last phase of my plan

"Phase 4 going to be the grand finale so to speak. After we have all of the prisoners in our possession the last step is getting them, as well as ourselves, out of there in one piece. Thanks to the explosives that Tank will have set up around the base, this shouldn't be a big problem. Once both transports are clear of the blast I will signal him to blow the charges. Thus sending the base into a state of panic, allowing us to slip out through the front gates virtually unnoticed. In the event that we are seen, the front transport; i.e. the one Lopez and I worked on, is equipped with weapons to deal with any resistance that we meet on the way out. Now, once we get out of the base we are going to need to go somewhere discrete to hide out. Despite all the secrecy that we have tried to create about our base here it's almost sure to be the first place that the EDF will check once they figure out what has happened."

"Well where do you suggest we go Stephen?" my father asks "There aren't exactly many places for us to go on the planet."

"Do you remember that place you took me to years ago when you were teaching me to drive?"

"You mean that old E-126 mining facility northeast of town?"

"The very same" I exclaim "It's the perfect place to lie low at for a little while. The way I see it considering it's been all but forgotten by the general population and EDF alike. Plus if they do ever think to send a drone over that way there is plenty of cover we can use to hide our activities."

"I like it… Alright is that all you have to say Stephen?"

"Yep, other than to say that this is one of the ballsiest plans that has ever been concocted in an effort to spite the EDF, and it will take every single one of us to pull off." I wave my father forward allowing him to take the stage.

"Thank you Stephen, alright you all have your assignments. Our family and friends were taking from us. Now, it's time we took them back. Who's with me!?" The small crowd erupts in cheers and fists gravitate towards the ceiling of the cave. Everyone begins to pile into the van, every other person grabbing an ICD as they entered taking their seats on the benches that were fastened to the cold steel that protected us from the planet outside. The ATMV Lurches into motion causing our bodies to sway to and fro. We sit in silence as we await the signal to cover up. I looked from face to face as we bumped along the road towards our destination when I realize something for the first time; If anything went wrong, people were going to die. Everyone else seemed to have come to grips with this thought; accepting it as the only possible way to end this conflict and secure their rights as Martian citizens. I just need everything to go off without a hitch that way I don't have to worry about the possibility of there being some sort of conflict. Wishful thinking I suppose…

12 OPERATION: TROJAN MARTIAN

My father bangs on the front facing wall of the cabin signaling that the time has come to cover up. I Slide the ICD over Cloee and me, activating it as it flows very softly over the both of us. We sit in silence then I feel her hand find mine in the darkness. She wraps her fingers around mine and they become intertwined. Sit in the darkness like this until she taps my shoulder, I slowly turn to face her she says something I can hardly hear, grabs my face pulling it to her lips. I feel their warming presence against mine; my free hand almost instinctively cradles her head. She slowly pulls away from my lips "Just in case something happens I just wanted to know that I really like you, and this last week has been absolutely delightful. Thank you for bringing me along on this journey I just hope it doesn't end before it ever truly begins," I start to open my mouth the respond when I feel our van come to a complete stop closing it back up almost immediately.

I hear muffled voices through the small slit left so that the drivers can still communicate with the prisoners and make sure they aren't harming each other. The sound of rustling papers creeps its way into the back of van then silence overwhelms us. Cloee finds my hand again in the darkness we lock our fingers together each squeezing the others. A loud bang against the rear of the van startles me but I do my best to remain as still as silent as possible. Shortly following the bang I hear the muffled voices start again and feel our ATMV inch forward. A feeling of relief flows across me , *We've made it through phase one; now onto phase two I just hope for their sakes they can get the job done.* It takes only a few minutes to move across the base but an eternity seemed to pass in between the entrance and the holding cells. The front doors of our van swing open with a noise similar to scratching a chalkboard, slamming shut moments later. Some of us are startled by the slam but shrug it off.

It only took a few minutes for Phase two to commence. Two quick hits on the rear door of the ATMV signaled we were a go. We all shed the ICD's leaving them on the floor of the van, the doors of the van flung open showing my father standing in an EDF uniform waving to us to hurry up. Quickly piling out of the van we rush into the holding area. I quickly spot the two EDF escorts a man and a woman; both lie motionless in the corner thanks no doubt to a good rifle butt to the head. I call everyone together briefly to clarify our plans

"Job well done Dad, and Lopez but this party isn't over yet. Now for this next part everyone needs to move fast. Tank you're on explosives; Cloee and I will go for the additional EDF transport, Lilly you need to work on freeing the prisoners and Cornelius and Lopez you two need to guard Lilly as she gets all of those prisoners into the transports. Understood?" Everyone quickly nods knowing time is not on our side

"Good now get to it!" The small group disperses to their various assignments Cloee and I quickly undress and slip into the escort officer's outfits. As a side note she was just as beautiful without clothes on as I had imagined. Once suited up we each grabbed a rifle. With me leading the way I grasped the luke warm knob of the door when my father's hand clasped onto my shoulder. "Be careful out there son, I don't want to lose you again." "Don't worry dad, I'll be perfectly fine." I wrap my hands around his body briefly pulling him in for an embrace. I turned the knob pushing the door out, allowing the light of the sun to breach into the room highlighting the particles of dust as it floats carelessly through the air. We slid through the door throwing down our visors shading the base and landscape in a black tint. Walking as casual as the two of us could we arrived at the vehicle bay; we were approaching the building when two men began making their way over to us shouting for us to stop.

Breaking into a jog when we stop moving the two men block our path.

"Gary, Steph, thought you two were supposed to be watching over that prisoner transport taking place right now?"
"Yeah we were but they….uhh they were uhhh…." I start to choke on my own spit failing to complete my thought out of fear brought upon by these men.

"Everything alright man?"

"Yeah your voice sounds pretty weird." Both men appear to be catching onto us *I have to act quickly otherwise the whole plan is a bust.*

"No guys I'm fine I just had some of that Martian Surprise from the mess hall today. The Surprise is it tastes like shit and kills your voice." They both look to each other for a brief moment then turn their attention back to me.

"Ew you actually had that?! How aren't you blowing chunks right now man?" We all start laughing trailing off after what just seemed to be a few seconds.

"So why aren't you two over at the Holding cells again?" the man on the right asks as he approaches me getting within inches of my face. I open my mouth to respond to him but Cloee beats me to it.

"We are getting a Second transport, due to the high volume of Prisoners being moved. It was an order straight from Wilkins himself in fact!"

"Oh," he says backing away "Sorry for the trouble, see you round Gary." I shoot Cloee a quick glance to show my approval of her handling of the situation.

Making our way inside the vehicle bay we quickly selected an armored ATMV with mounted Light machine gun finding the keys hanging on the wall. Just in case the EDF happened upon our plan and intend to follow us I pulled a knife and punctured the tires of all ATMV's that were in the Vehicle bay. I quickly climbed into the driver seat while Cloee climbed into the passenger seat buckling her seatbelt as soon as she was settled. I slid the keys into the ignition and the ATMV roared into life. Careful not to cause to anything I slowly pulled around to the Holding cell.

"Stay here, I'll let them know we are good to go." I tell Cloee slipping out onto the ground.

"Alright I'll be here." I causally jog over to the holding cells, opening the door again allowing the light from the outside to penetrate into the room. However, I don't hear anyone inside all I see is darkness. I step inside whispering "Guys you here? It's just me Stephen….Dad….Lopez….Lilly…. Anyone? Hello?"

"STEPHEN IT'S A TR-" Cloee warns from outside but it's too late, five EDF soldiers have me pinned in mere seconds and commence to beat the ever leaving hell out of me; Dragging me outside after they were satisfied with the beating they had administered to me, they threw me into a line of rebels with Cloee not far behind me. I glance around me to try to recognize any of the faces that surround me, slowly coming to the realization that we all have been put on the firing line. My vision clears just enough to see the group of people that had been gathered in front of us. I notice on the ground that a shadow is slowly creeping its way back and forth in front of us, looking upward to see who owned the shadow I can just make out Commander Wilkins ugly mug as strolls back and forth. I look down the line to make sure that all six of us are accounted for; thankfully we all are. I lock my focus on Tank who has felt my eyes on him and looks my direction. I nod my head bringing a smile across his face just as Wilkins reaches him. Noticing the smile he stops walking and confronts him "Why the heck are you smiling little man, your plan was a failure! All of your friends and neighbors are still in my custody. And I plan to kill all of you. So what reason do you have to be happy?"

Tank very softly responds "Because I get to watch this base burn to the ground with you in it." He reveals the detonator from his pocket, flipping the cover away from the button pressing it down with his finger and swinging a clean right uppercut into the Generals crotch; all the while Explosions erupt all around us. *Well, So much for in and out without a sound. Looks like we are going to have to go to plan B...kill every son of a bitch that gets in between us, and that gate*

13 THE PLAN GOES SOUTH

Following tanks lead we all jumped up and went back to work on the initial plan. My father runs over to the crowd shouting at them "Get inside the transport now!" frantically they scramble towards the ATMV while the sound of automatic weapons fill the air. Pinned down behind cover hearing the constant ping of bullets ricocheting off the pieces of metal I was using as cover. The assault rifle I had stolen swung down from my shoulder unfired. Glancing to my right I see Lilly huddled over behind a barrel, she raises her head slightly spotting me. Raising her palm towards me retracting finger after finger, I begin to catch on to what she is doing with her palm; she's counting down, *but what does she want me to do when she reaches zero? Does she want me to join her over there?* Three fingers still stand. *Does she want me to return fire on the EDF?* Two fingers still stand. *Does she perhaps want me to cover her as she makes a break for the transport?* A lone finger still stands. She puts herself into a position to run towards me, well looks like she wants some cover while she makes a break for it. The last finger falls. I signal to her that I'm ready to go. She breaks into a full out sprint for my position, unclicking the safety on the gun I spin around the corner and start to unload into the onslaught of EDF coming towards us.

Time seemed to slow down to a crawl as Lilly made her way towards me. I feel every time my gun kicked back against my shoulder, I watch as each EDF trooper I hit reels backward falling onto the ground having a pool of their own blood form around them. I see every bullet as it fly's through the air; almost as if I can reach out and pluck it from its path. Grenades bring time back to normal. Several EDF pull grenades from their belts lobbing them towards Lilly. The first two do nothing to slow her run but the third is too close for her just to ignore. She is flung to her side from the explosion. I am knocked onto my back as the sound of the explosion resonates in my ears causing them to ring. I lay there on my back staring at the sky thinking *well, this is it. This is where I die. Getting blown up by an EDF Grenade…not my choice of death but hey when you're dead you're dead…*I slowly turn my head to the left, seeing my father quickly funnel people into the transport van that only moments ago would have helped me escape this hell hole and return back to my home. I see someone who looks similar to Cloee, no wait that is Cloee, making her way to me with Tank not too far behind her. I very weakly am able to turn to my right; Where Lilly lay motionless in a growing puddle of red. Her eyes remain wide open but completely devoid of life. The same pair of eyes that less than five minutes ago were full of drive and determination now lies cold and lifeless. I can't bear to look at her any longer I turn my head to face the sky again.

She's dead…those bastards killed her and it's all my fault… tears begin streaming down my face as Cloee and tank both arrive. Both seem extremely panicked, tank motions to Cloee to get me and then ran out of my view. Cloee stoops down and pulls me up off my back. She is shouting something at me but my ears still ring too bad for me to make sense of anything she is trying to say. Eventually she gives up helping me to my feet putting my arm around her shoulder walking me, well really more carrying me, to the Transport van. When we arrive I have regained the majority of feeling throughout my body, and can walk under my own power. Cloee goes to help me into the back of the transport but I resist "No, I've got to get some payback on these bastards for lily. Plus I am the only one who knows how to operate the turret." She opens her mouth then decides it's not worth fighting me over; instead deciding to kiss me, mouthing "I love you." I mouth back "I Love you too." She climbs into the back of the van after tank appears with Lilly's dead body carefully entering the van as well. Once both are in I close the doors and hobble as fast as I can to the Armored ATMV Cloee and I had stolen earlier. My father gave me cover as I made my way to the vehicle. Once I reach it he whistles to Lopez signaling that it is time go. He climbs into the Transport van while my father and I climb into the armored car; I situate myself in front of the onboard turret controls. There is the sound of constant collision of EDF bullets against the metal of the armored vehicle. Dad turns the keys and the ATMV roars into life more or less peeling out as he drives forward with the turret coming to life not long after. I rotate the turret from side to side firing at anyone unlucky enough to try to stop us.

I must have shot and killed at least 37 Soldiers while we made our escape. Once we had shoot our way out of the camp we made a break for the exit; Where, faced with the challenged of a heavily fortified gate. "Better hold on tight junior." My father shouts back "This is gonna be one hell of an impact." The engine revs tilting the frame of the ATMV upwards slightly. I quickly throw myself into a seat and strap in. The force of the impact almost rips me free of my restraints and brings the ATMV to a complete stop. Everything that wasn't nailed down or securely attached flies forward. My head whips forward than backwards against the seat. My head throbs from the impact and I struggle to keep vision straight but it quickly becomes hazy. I fumble to un-buckle myself from my restraint eventually releasing my body to the planets gravitational pull. I collide with the ground once again slamming my head against the cold floor; everything begins to be enveloped in darkness. With the last of my strength I lift my head to see my father climbing backwards towards me he just reaches me when everything goes black.

14 THE INTERROGATION

I remember waking up in a daze alone in a dusty room, wallpaper curls down the walls on all sides. I try to lift my head but it proves to be an impossible task. *Oh god I am paralyzed?* I attempt to clench my fists, and then tried to move my feet but no luck. My heart beat escalates to an irregular speed causing the heart monitor attached to me to go ballistic. I hear the creak of an old door crash open, and the sound of footsteps colliding with the metal floor. They rush to lower my heartbeat; one of them stands over my head holding a needle. The person descends on my quivering arm. *Wait, my arms quivering! That means I'm not paralyzed…at least not entirely. It must be something with this table.* I feel the needle penetrate my arm releasing a numbing sensation into my blood stream; that seemed to spread like a virus reaching all over my body in just matter of seconds. The heart monitor begins to quiet down returning to a normal tempo. Suddenly my eyes grow heavy and my vision starts to become fuzzy again. My eyes drift closed and I fall into a deep sleep.

I woke up after some time free of the restraints that had caused my heart rate to spike. In an effort to find out for sure where I was. I started to lift my head, with my body trying to follow suit but pain restricted me and forced me to my back again. *Holy crap that hurt* I think to myself while gripping my ribs. I try two more times to pick myself up to similar results. Eventually I just gave up and lay on the table. A few minutes passed and I hear the door creak open stealing my attention from the ceiling; it was Cloee. She looked as beautiful as I had ever remembered as she walked across the room to me. She knelt down beside me laying her hand on top of mine staring deeply into my eyes.

"Thank god you're ok" she said finally "'Ok' must be a relative term then cause I feel like shit." I say with a chuckle and an immediate wince caused by the pain. She leans down, closes her eyes and delicately places her lips on mine. "Feel better?" she says straightening herself up

"A little" I say winking at her "Be even better if I could get off this table though."

"I don't know if you should. Your injuries were pretty severe." She stands up and walks across the room to a metal desk that looked like it should be in a heap somewhere twisted and mangled among other decrepit equipment. She pulls open a drawer grabbing a manila folder from it and slowly making her way back to me. She grabs a chair and places it next to me. Opening the folder she tells me all of the injuries I sustained as a result of the high speed collision into the gate. My injuries included but were not limited to: a concussion, whip lash, seatbelt burns, and broken ribs. All in all an impressive injury report for my first accident. "Well that must have been quite the crash to cause that many injuries." I say with a smirk running across my face. Cloee smiles "Yes it is, but let's not try to beat it okay?"

"Fine, I promise…Hey where is everyone else?"

"Who do you mean by everyone?" this struck me as strange, how couldn't she know who I was talking about but I decided it was just poor memory; after all she did just take part in a huge battle. "You know, Lopez, Dad, Tank, the people we managed to free, and Lilly." I choke… 'Lilly' the last time I saw her she was carried into the back of the transport by Tank. I turned to Cloee as I awaited her answer. From the light above my table I could see tears begin to creep down her face "Everyone's dead Stephen."

"What?" I respond softly

"Everyone is dead… They all died in the crash."

"No, that's impossible my father was alive after that crash! I saw him with my own eyes climbing back to get me. We opened the gate.…" I trail off as tears begin streaming down my face. *Lilly, Tank, Lopez, Mom, Dad… all dead… the only people left are Cloee and I. Wait,* **Cloee** *and I.* "Wait a minute, how are the two of us still alive if everyone else is dead? Where are w-?" The door swings open allowing natural light to seep into the room; her face goes pale as she lays eyes on the figure in the doorway. I turn my head towards the door way and When my eyes finally met with the figure standing in the door I realized why Cloee had lost color in her face. The figure standing in the doorway was General Wilkins himself.

15 JAILBREAK

"Well we meet again Mr. Stephen Crow. That is your name correct?" He looked like hell close up. It looked like he hadn't shaved in weeks or slept for that matter either. The bags under his eyes and the gray hair on his head told his struggle with being a general. "Damn it Wilkins I thought I told you I had this under control!" it was Cloee voice. I turn to her with a look of confusion. She simply looked at me and shrugged "Well it looks like the cats out of the bag, no need for this…" moving her hands behind her ears she pulled forward deactivating her disguise. What was revealed from behind the mask was somewhat of a surprise. An absolutely breath taking woman. She let loose her long dark hair that stopped just below her shoulders. A black spandex jumpsuit with hexagonal patterns hugged her body. I caught a glimpse of her name embroidered just above her right breast: *Claire*

"I was just about to get the location from him before you came and ruined everything." She said turning her attention to the General.

"You were getting nowhere, if anything you were the one who compromised the interrogation." He replied. The pair stood arguing for several minutes completely absorbed in what they were screaming at each other about. Glancing to my left I noticed the door was open and that there was a clear shout out of here *if I can just make it to the door I could get out of this place…where ever the hell this place is.* I lift my head up and look at the general and Claire, both are still screaming at the top of their lungs about me and the other rebels.

As quietly as I possibly could, I moved my legs to the left side of my table. One by one sliding them off of the table until I am sitting upright. The door sits in front of me wide open practically calling to me *Come on Stephen be free! All you have to do is run for it!* I grip the table hoisting myself into position to lunge for the door when an explosion erupts shaking the ground and everything attached to it as well as sending me tumbling onto the ground. The General and Claire are both sent staggering from the force of the blast. Once they regain their footing they rush to the door. *Damn it!* I think to myself, *there is no way I am gonna be able get out of here with them blocking the door.* Almost as if someone had read my mind, a metal baseball bat meets the forehead of Claire knocking her unconscious. In response the general revealed his pistol but it was too late the bat had found his head and landed cleanly on his skull. Bringing the almighty General Wilkins to his knees and finally to the ground in a lifeless heap. From the ground I couldn't make out exactly who the figure was. But when they bent over to help me up the light from the table caught on his big hand; it was tank's hand.

He lifted me up and slung my arm over his shoulder. "Come on little crow." He said "Can't have you lying down while we all worry about you." **We** I thought to myself *that means everyone is alive!* As we emerged from the bunker I was being held in the light blinded me. But I knew there was no time to allow my eyes to adjust. We ran across the base to a section of fencing with a large hole in it. I removed my arm from around tank so I could make it through the hole

"Sure you can walk kid?"

"Is the Sun hot?" I replied sarcastically invoking a smile to spread across his face.

I quickly climbed through the hole with tank following close behind. I let him pass in front of me because he obviously knew where we were going. We reached his ATMV which stuttered into life as he turned the key. The tires spun as he slammed the gas pedal down, wrenching the steering wheel to the right as we pulled onto the road. I peeked behind us and it seemed that the EDF were making no efforts to try to follow us. He gradually slowed down as he realized that we weren't being followed, and turned off the main road driving into what seemed like the middle of nowhere. I eventually noticed a large cliff that lay directly in front us, and it didn't seem tank had any intention of slowing down. I started to cling to my arm rests as our distance between the cliffs rapidly shrank. "Shouldn't we be slowing down to avoid this? Instead of hurtling full speed towards it?" Tank didn't respond, simply smirking and pressing the gas pedal even further down. I close my eyes and let loose a scream as we are inches away from the cliff. I soon realized that we were perfectly fine and stopped screaming *they must have figured out a way to amplify a PCD to work for a large area* I thought. Tank was beside himself. "HA you should have seen your face! Gah that was priceless" I slugged him in the shoulder and smiled "Don't ever do that again." We reached the old mining site from my original plan, where the faces of the rebels lite up as they saw the ATMV casually pulling through the streets. Soon we arrived at what must have been the Comm center when this Site was fully operational. I stepped out of the ATMV giving a nod of thanks to Tank for the rescue. One foot after another I made my way to the door of the building, slowly turning the knob and pushing the door in. I locked eyes with my father and mother who were quietly sobbing over what they must have assumed as the loss of their oldest son. I smile and say "I'm home."

www.ingramcontent.com/pod-product-compliance
Lightning Source LLC
Chambersburg PA
CBHW021414170526
45164CB00002B/644